Fossils, minerals and rocks

Collection and preservation

Fossils, minerals and rocks

Collection and preservation

R Croucher & A R Woolley

British Museum (Natural History)
Cambridge University Press

Acknowledgements

We gratefully acknowledge the help given by many of our colleagues at the British Museum (Natural History). Especially we would like to thank C. P. Palmer for field notes, sections etc., and Mrs E. G. Zielinska-Millar for the photographic work. Permission to reproduce figure 14 was kindly given by Imitor Ltd.

Published by the British Museum (Natural History), London
and the Press Syndicate of the University of Cambridge
The Pitt Building, Trumpington Street, Cambridge CB2 1RP
32 East 57th Street, New York, NY 10022, USA
296 Beaconsfield Parade, Middle Park, Melbourne 3206, Australia

First published 1982

Printed in Great Britain by Henry Ling Ltd,
Dorchester, Dorset

ISBN 0 521 24736 5 hard covers

Library of Congress Cataloguing in Publication Data: 82–1282

British Library Cataloguing in Publication Data
Croucher, R.
Fossils, minerals and rocks: collection and preservation.
1. Palaeontology–Collecting of specimens
2. Geological specimens–Collection and preservation
I. Title II. Woolley, A. R.
560 QE718

Contents

7 **Introduction**
7 *Conservation*
8 *Safety*
10 **Field equipment**
10 *Clothing*
10 *Tools*
15 **Sources of information**
15 *Guides and memoirs*
16 *Journals*
16 *Books*
17 *Maps*
18 **Minerals and rocks**
18 *Collecting*
19 *Recording*
21 *Transport*
22 *Preservation of specimens*
23 *Curating a collection*
27 **Fossils**
27 *Collecting*
27 Recording
28 Extraction
39 *Preparation (Development)*
39 Mechanical
44 Chemical
45 *Conservation*
45 Preparation
45 Consolidation
46 Repair
48 *Pyrite conservation*
49 *Curation*
50 *Chemical hazards*
53 **Further reading and sources of reference**
54 **Materials, manufacturers and suppliers**
57 **Addresses**
59 **Index**

Introduction

In recent years there has been an enormous increase in interest in fossils, minerals, and rocks, that has been catered for by a variety of books and magazines on the subject. Nevertheless, some aspects have been neglected, notably, techniques for making and conserving a collection. This short guide endeavours to fill this gap, and it is intended both for the beginner and for the experienced collector who requires information on some specialized techniques, particularly those of collecting and preserving fossil vertebrates.

Conservation

In a booklet concerned with collecting of fossils, minerals, and rocks, the importance of conservation must be stressed at the outset, especially since indiscriminate collecting is destroying many of the better known and finer collecting localities. In the British Isles the literal demolition by collectors of some of these localities has provoked the 'Landed Farmer's Association' and other organizations to try to stop collecting from certain sites, particularly by organized parties. The Geologists' Association have published *A Code for Geological Field Work* which should be observed, while the Nature Conservancy is also actively engaged in listing important sites and trying to protect them, not only from the enthusiastic collector but also from being spoiled or destroyed by other means – see the booklet *Conservation and the earth sciences* published by the Nature Conservancy Council. Some sites are now effectively protected and can only be worked for scientific purposes.

In the interests of conservation, mineral veins, and strata particularly rich in fossils, should not be hammered if this can be avoided, for there is usually loose material which is both adequate and more easily collected.

Permission should always be sought to visit sites on private land and particularly farm land, and care taken to shut gates, not to frighten livestock, and not to scatter rock debris. Good material can often be obtained from working quarries, but these can be dangerous and prior permission for access must be obtained. In recent years the motorway building programme has revealed many fine rock sections which for the fossil collector in particular may be very attractive, especially as they are often in areas otherwise poorly exposed. There is an obvious danger from earth-moving plant and collapse of sections when working in such areas, and it is essential that permission from the relevant authority should be sought before attempting to visit them.

Safety

The vigorous hammering of rocks, particularly tough ones, can be dangerous and a number of simple precautions should be taken. Use only a geological hammer because these have specially tempered heads; other hammers are likely to splinter and result in injury, particularly to the eyes. Special care should be taken if it is necessary to hammer at eye level, for

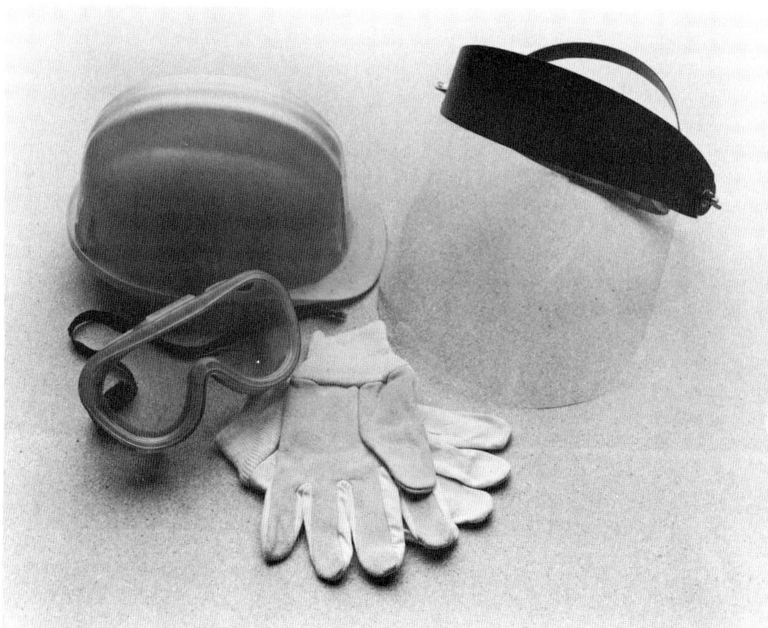

Fig. 1 Basic safety equipment for field and laboratory use: goggles, visor, safety helmet and stout gloves

instance on cliff or quarry faces, and in the vicinity of other collectors. When hammering, plastic goggles should always be worn (fig. 1). Never attempt to split rocks by inserting the chisel edge of one hammer into a crack and then hitting this hammer with a second; a misdirected blow could project jagged metal splinters at a considerable velocity. Always use a steel chisel. Cliffs, like quarries, can be very fruitful collecting sites, but are dangerous places. Apart from the obvious care needed if it proves necessary to climb, there is a constant danger from falling rocks. It is advisable in such places to wear a safety helmet (fig. 1), and nowadays in working quarries it is a prerequisite that helmets should be worn before the company owning the quarry will allow visits.

One of the most attractive features of fossil, mineral and rock collecting, is that it gets one out into the open air, often into remote and challenging country. In such circumstances, especially if collecting alone, it is necessary to indicate to someone your intended destination and route, particularly in mountainous areas.

Field equipment

Clothing

Provision should be made for the changeability of the British climate, particularly in mountainous areas. Thus, in addition to durable clothes, spare warm clothing and waterproofs should be carried; and stout footwear is essential.

Tools

The tools most commonly required by fossil, mineral and rock collectors (figs 2 and 3) are: geological hammers, sledge hammers, cold chisels, wedges, crowbars, pick-axe, trowels, shovel, sieves, knife and lens.

A range of geological hammers is illustrated in Figure 2.

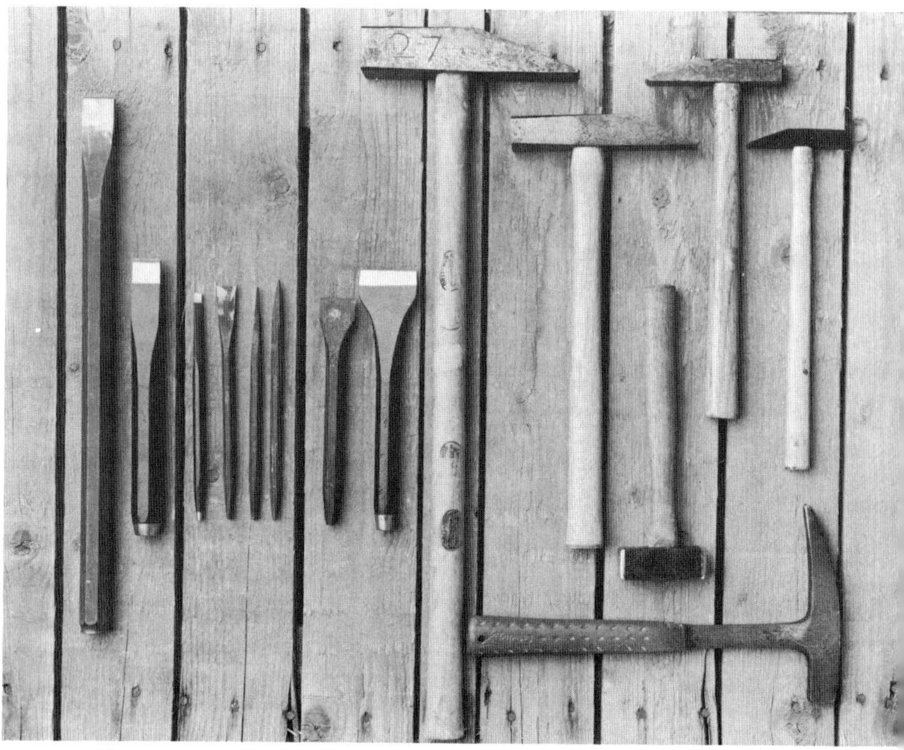

Fig. 2 A range of geological hammers from 120 g (4 oz) to 1·1 kg ($2\frac{1}{2}$ lb); a club hammer; and cold chisels, from points to bolsters

Fig. 3 A selection of commonly used field tools and materials including sledge and club hammers, pick and mattock, both with detachable handles, trowels, knives, brushes, scissors, shovel, sieves, disposable p.v.c. gloves, scrim and tissue-paper

Most have a square head and an opposing chisel edge (or wedge), though hammers with a pick end (fig. 2) are also available, and they range in weight from about a hundred grammes (for trimming) up to about two kilogrammes. For hard rocks a one-kilogramme hammer is most universally used, but a lighter one is available for softer rocks. Hammer handles are usually of wood, generally hickory, but glassfibre handles and all-steel hammers can also be obtained.

Sledge hammers also range in weight, but those between two and five kilogrammes are most satisfactory. They are required for collecting large specimens, and for working tough rocks.

Many rocks are so tough or form such smooth, rounded outcrops that they resist even the most intense hammering; but few can withstand the pressures applied by a cold chisel (fig. 2) hammered into a judiciously chosen crack. A chisel is even more important when accuracy is required. Thus, a fine mineral specimen may easily be damaged by a mis-directed hammer blow, but usually some crack or joint can be found into which a chisel can be worked with little danger to the specimen. Similarly, when collecting fossils it is often difficult to predict which way the rock will break by hammering, but careful use of a chisel will provide greater control. Cold chisels are made from forged and heat-treated carbon steel or chrome alloy steel, so that they will not fracture on hammering, though it is usual for the metal at the butt end to curl over with time. This metal should be carefully ground away before it breaks as metallic splinters. Chisels vary in their lengths and in the width of the cutting edge. For more precise work and for removing tiny specimens, a 5-mm wide blade or less, or even a fine 'point' or 'pin' chisel may be used, but for general use a 25-mm blade can be recommended.

Wedges of wood or metal can sometimes usefully be employed in conjunction with a chisel. Cracks, laminae or joints when opened and widened with a chisel, frequently close again when the chisel is removed. In such circumstances they may be kept open with one or more wedges, and then extended by further chiselling. This technique is valuable for removing large slabs of sedimentary rock, though crowbars and picks are also useful for this purpose as well as for enlarging exposures.

Many fossils are obtained from soft rocks by sieving of bulk

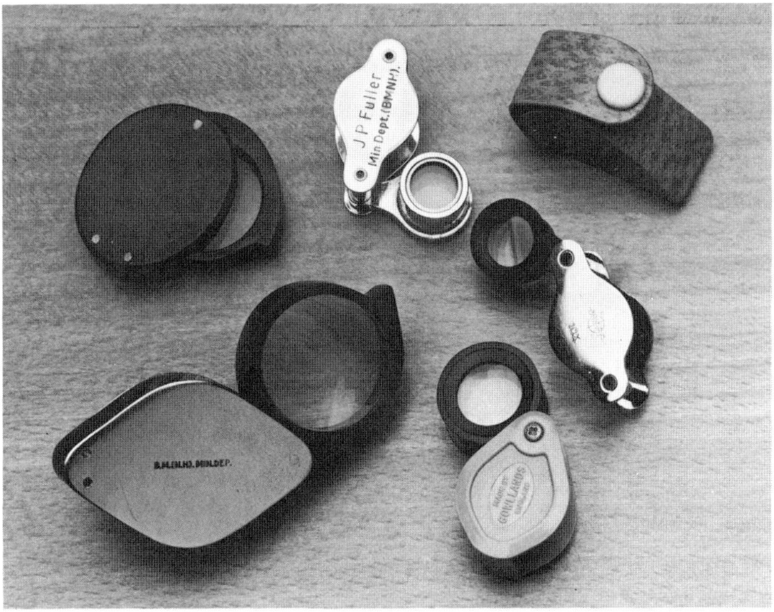

Fig. 4 Hand lenses suitable for use in the field or laboratory. The handle folds to protect the lens

samples, either in the field or the laboratory. If done in the field, the sieve (fig. 3) should not be coarser than 0·5 mm gauge (500 mesh) otherwise small fossils, such as teeth, may easily be lost. A stack of sieves may be used to separate different sized fractions. Dry sand is easily handled, but many sediments need to be washed through the sieves with water. Clays are best treated in the laboratory where they are dried and then wet-sieved using hot water.

Trowels, preferably the diamond-shaped ‘pointing’ pattern (fig. 3) are used for removing fossils from relatively soft strata. A knife, such as a sheath knife (fig. 3), can be used in the same way, as well as for cutting out blocks of sediment containing fragile fossils.

A hand lens is indispensable to collectors of natural history specimens of any kind, and a range of suitable types and magnifications can be purchased at moderate prices. The most useful are those that fold back into the handle, which protects the glass when not in use (fig. 4). A magnification of × 5 to × 15 is convenient. Such lenses are best used by holding them close to the eye with one hand while the specimen to be

Fig. 5 A hand lens should be used by holding it close to the eye and moving the specimen into focus

examined is brought into focus with the other (fig. 5). The technique is to move the specimen and not the lens.

A pocket knife with a sharp blade may be used for prising small crystals or fossils from the matrix, and a high quality blade can be used for scratching minerals when testing hardness as an aid in identification. For the identification of carbonate-bearing minerals and rocks, some collectors carry a bottle of dilute hydrochloric acid. A small plastic bottle, preferably with a narrow nozzle for the application of a single drop, is safest; the acid must be dilute.

Other essential items of equipment are a notebook, pencil and pen, a plentiful supply of wrapping materials, and a rucksack or bag for carrying equipment and specimens. A camera is also desirable. The use of these items will be detailed later.

Sources of information

It is neither practicable nor desirable to list collecting sites. Instead the collector must familiarize himself with the geological literature from which information can be obtained on the range, location, and nature of good collecting localities. Much information can be found in 'guides' and 'journals' which are relatively cheap to purchase, and may be held by a good general library, but it may be necessary to consult geological memoirs and maps which are to be found only in specialized geological libraries. Such libraries are often held by geological societies, and joining such a society is the easiest way of gaining access to journals, books, and geological maps. Such a society is the Geologists' Association, which caters both for the amateur and the professional, and has regional groups in different parts of the country. Field excursions are organized, some especially for the beginner, and field guides are published. Further information can be obtained from: The General Secretary, The Geologists' Association, Burlington House, Piccadilly, London W1V 0JU.

Guides and memoirs

The most comprehensive set of guides available are those published by the Geologists' Association (obtainable from the address given above, or from the bookstall of the Geological Museum, Exhibition Road, London SW7 2DE). These cover areas throughout the United Kingdom.

There are many other guides for particular areas of Britain, many of them published by local geological societies. Full lists of these guides can be found in the following two references:

Bassett, D. A. 1969. Directory of British Geology 1, A provisional annotated bibliography and index of geological excursion guides and reports for areas in Britain, C, England. *Welsh Geological Quarterly,* **3** (3/4), 3–91.

Sargeant, W. A. S. & Harvey, A. P. 1973. Stratigraphy (historical geology) including regional geology. In *Use of earth science literature* edited by D. N. Wood, London, Butterworth.

For general background geological information and covering all aspects of geology, including economic geology and palaeontology, the 17 very well illustrated handbooks on the *Regional Geology of Great Britain* are very good value. These are obtainable from HMSO bookshops or the Geological Museum, Exhibition Road, London SW7 2DE, and other bookshops.

The detailed geology of the United Kingdom in all its aspects is described in the numerous memoirs of the Institute of Geological Sciences. Lists of these memoirs are available from HMSO bookshops and the Geological Museum, and the memoirs relevant to particular areas are listed in the Regional Geology handbooks referred to above. Many memoirs are out of print but can be consulted in libraries.

Journals

Journals range from those which are devoted to the publication of research papers for the professional and the most dedicated amateurs to more general ones specifically for the amateur. A selection of these journals follows:

Proceedings of the Geologists' Association – partly for the professional, but many general papers also.

Journal of the Geological Society – essentially a research journal.

The Geological Curator (formerly the *GCG Newsletter*) – journal of the Geological Curators Group (affiliated to the Geological Society).

Geological Magazine – essentially a research journal.

Mineralogical Magazine – essentially a research journal.

The Journal of Gemmology – mostly professional but some papers on collecting localities etc.

Palaeontology – essentially a research journal.

Gems. The British Lapidary and Mineral Magazine –caters specifically for the amateur: enquiries to The Randal Press Ltd, 9 Kennet Road, Crayford, Kent.

Gem Craft – for the amateur: write to *Gem Craft*, Subscription Dept., PO Box 35, Bridge Street, Hemel Hempstead, Herts, HP1 1EE.

Books

The range of books in geology, palaeontology and mineralogy

is vast and in recent years there has been a particularly large output of publications aimed at the amateur, the more popular of which are obtainable at good bookshops. Rather than list a selection of these it is suggested that the interested reader should browse over the shelves of his local library and bookshops to discover for himself what is available along the lines of his own interest. For the fossil collector however, three publications of the British Museum (Natural History) *British Palaeozoic Fossils, British Mesozoic Fossils,* and *British Caenozoic Fossils* are well worth acquiring, as they have illustrations of many of the more common fossils to be found in the British Isles.

Maps

Both topographical and geological maps are an essential part of the equipment of the serious collector. The 'standard' topographical maps for most purposes are the 1 : 50 000 (which have replaced the 1 inch to 1 mile) sheets of the Ordnance Survey. More detailed maps on the scale of 1 : 10 000 (the older 6 inches to 1 mile may still be available) are available for most of Britain and for some areas 25 inches to 1 mile.

In the United Kingdom the most detailed geological maps generally available are the 1 inch to 1 mile series (being replaced by 1 : 50 000) published by the Institute of Geological Sciences. These sheets have been published for most of the UK but many sheets are out of print and must be consulted in a library. A most useful series on a scale of 4 miles to 1 inch provides coverage for the whole of the UK, while two sheets on a scale of 10 miles to the inch, and obtainable folded, cover the whole of Great Britain. All the above maps are printed in colour. Many of the larger scale maps are available in 'drift' (that is all the superficial deposits are shown) and 'solid' (all the superficial deposits are removed) editions.

All these geological maps are available from the Geological Museum, Exhibition Road, London SW7 2DE or through HMSO bookshops from whom lists of the maps available can be obtained.

Minerals and rocks

Collecting

The greatest difficulty when collecting minerals, particularly if they are coarsely crystallized, is their removal from a vein or cavity without damage. It is best to try to remove the crystals together with the rock (matrix) to which they are attached. This should be done with a hammer and chisel, not a hammer alone, utilizing cracks, bedding planes, or other directions along which the rock is likely to split. Resist the temptation to trim away the matrix, unless you are absolutely sure of the way it is likely to break. It is much better to take away the whole piece and to trim it at leisure, than to ruin a fine specimen through undue haste.

Collecting material on matrix has the advantage of protecting the specimen, and making it less likely to break in transit. Further, display specimens often look much more attractive on matrix than as single crystals.

Some minerals have perfect cleavages along which they break readily. Calcite is perhaps the best known such mineral, and even if a crystal is not broken by the attempt to remove it from matrix, incipient cleavages may be opened and their presence spoils the appearance of a clear crystal. Minerals such as fine-grained, powdery coatings, known as 'blooms', and minerals which have developed along joint and bedding planes such as dendritic growths of manganese oxides, are almost impossible to separate from matrix.

The only major constraint when collecting rock specimens is to find material which is 'fresh', i.e. unweathered. Working quarries expose fresh surfaces from which good material can be collected; but in natural exposures the surface is usually weathered and the underlying fresh rock must be obtained by hammering.

Occasionally, there is an advantage in collecting weathered material. The different minerals in a rock may break down at different rates, with the result that one particular variety may stand proud of the surface as good clean crystals. With such material it is probably better to try to collect a specimen having one weathered and one fresh surface.

In the past it was usual to trim rock specimens into rectangular blocks of about $50 \times 80 \times 100$ mm, and many museum collections contain drawers of such neatly regimented specimens. Nowadays, specimens are rarely trimmed to a uniform size, but should be trimmed so as to remove sharp corners or weathered patches, and to get as close to the ideal rectangle as possible.

Recording

All collected mineral and rock specimens should be documented as fully as possible, recording their field relationships, the locality from which they came, and the date when collected (fig. 6).

The field relationships of a specimen are recorded in order to place it in its geological context, and with research material this may be crucial to any interpretation of how the mineral or rock was formed. If a specimen comes from a mine or quarry tip, little can be done except to note the locality. But for specimens collected *in situ*, whether naturally occurring or in a quarry or mine, one should note whether the specimen comes from a vein, a dyke, a lava flow, forms concretions, occurs in vugs, and so on. Recording is often facilitated by sketches and by photographs. The advantage of sketches is that emphasis can be given to the geological features that may not be obvious in a photograph. Photographs are, however, particularly valuable in recording well-developed structural features, and as an aid to memory. All this information should be written in a notebook, together with the locality, recorded as accurately as possible (fig. 6). In the British Isles the best method is to use a national grid reference. These data are used later for making catalogue entries and for labels. Specimens should be numbered or labelled in the field, the numbers being entered also against the information in a notebook. The choice of numbering system is immaterial provided specimens can later be referred to notebook entries. If localities are marked on a map, then the field number should also be shown to facilitate locality identification at a later date.

A variety of methods is used for numbering specimens. Some collectors prefer to write the number on a slip of paper which is wrapped in with the specimen, to be replaced by a permanent number later. This method is not recommended because loose paper slips tend to go astray. It is better to write

Fig. 6 Recording in the field is facilitated by the use of detailed maps and photographs, and by the careful recording in a notebook of relevant information including sketches of sections and other details

Fig. 7 Specimens may be numbered in the field in various ways such as writing directly on to the specimen with a felt tipped pen (right-hand specimen). More permanent numbering may be achieved with paper tickets (both specimens) or a dab of white paint on which the number may be written (left-hand specimen) and protected further, if necessary, by a coat of clear varnish. A permanent label should be kept with each specimen

directly on the specimen with a permanent felt tipped pen (fig. 7), or to write the number on a short length of adhesive, linen tape, which is then attached to the specimen. Writing directly on to the specimen is not acceptable with finer mineral specimens, but linen tape can be peeled off later, usually without doing much damage.

Transport

Although few rock specimens are fragile, they must be wrapped adequately in newspaper to prevent them rubbing together in transit. Such rubbing causes an unsightly white patina, and may obscure detail and generally scratch the surface. Polythene or linen bags can also be used (fig. 8). Many mineral specimens can also be carried safely after being

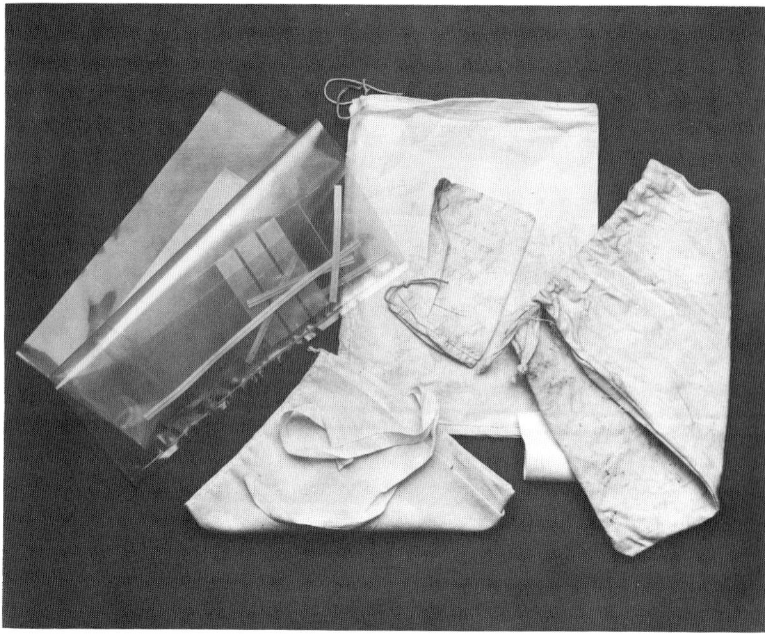

Fig. 8 A range of linen and polythene bags for carrying specimens in the field, although newspaper is generally adequate and probably affords greater protection

liberally wrapped in newspaper, but delicate specimens call for special care, and are best carried in wooden, cardboard, or metal boxes lined with some soft packing material. Cotton wool should not be used for mineral specimens which form fine needles, as the strands tend to get caught up between the needles and are difficult to remove without damage. Loose polystyrene chips can be effective, but they tend to shake to one side, leaving the specimen open to damage. For carrying specimens back from the field some form of canvas shoulder bag is usually used, but for heavier loads a framed rucksack carried well up on the back is the most comfortable and effective.

Preservation of mineral specimens

Some minerals are subject to alteration and decay and must be treated if they are to be preserved. One of the most common minerals to be affected in this way is pyrite. The causes of this decay and some remedies for pyritized fossils are detailed on page 48, and the same technique should be applied to the preservation of pyrite mineral specimens.

Some minerals are deliquescent, that is they absorb moisture from the atmosphere, and they can be completely spoiled if left unprotected. Minerals particularly prone to this effect are chlorides and sulphates, and they can be preserved by putting them in a glass jar sealed with paraffin wax or 'Vaseline' (petroleum jelly), or in a polythene bag which is 'fuse sealed' and contains a few grains of silica gel to absorb moisture already present.

Some minerals are altered by light so should be stored in a light-tight box or drawer. Common examples are sulphides of silver, realgar (which alters to orpiment), and topaz, some sherry-coloured varieties of which may become blue.

Curating a collection

If a collection of rocks or minerals is to have permanent value it must be properly labelled and catalogued. A system of cataloguing should be adopted in which the numbers fixed to the specimens must correspond exactly with the entry in the master catalogue, which lists such information as species, locality, and date collected (fig. 9). A corresponding label giving the salient information should also be drafted and kept with the specimen. The numbering of specimens should be done neatly and well, for a lost or unreadable number can render a specimen as good as useless. The method adopted in the British Museum (Natural History) is for the number to be written in black, waterproof ink on a small, oval, paper ticket measuring 14 mm × 9 mm, which is stuck on to some unobtrusive part of the specimen (fig. 10). The adhesive used is 'Seccotine' and a scrap of cotton wool is added to the dab of Seccotine, which makes for a stronger bond. Before the label is firmly pressed into place with a piece of cloth, it is soaked in water for a few minutes which makes it pliable, so that it moulds readily to the contours of the specimen and thus prevents corners lifting off. Seccotine is soluble in water so that labels can be removed and transferred to another part of the specimen if necessary. However, this must be borne in mind if rock or mineral specimens are cut using a water-based

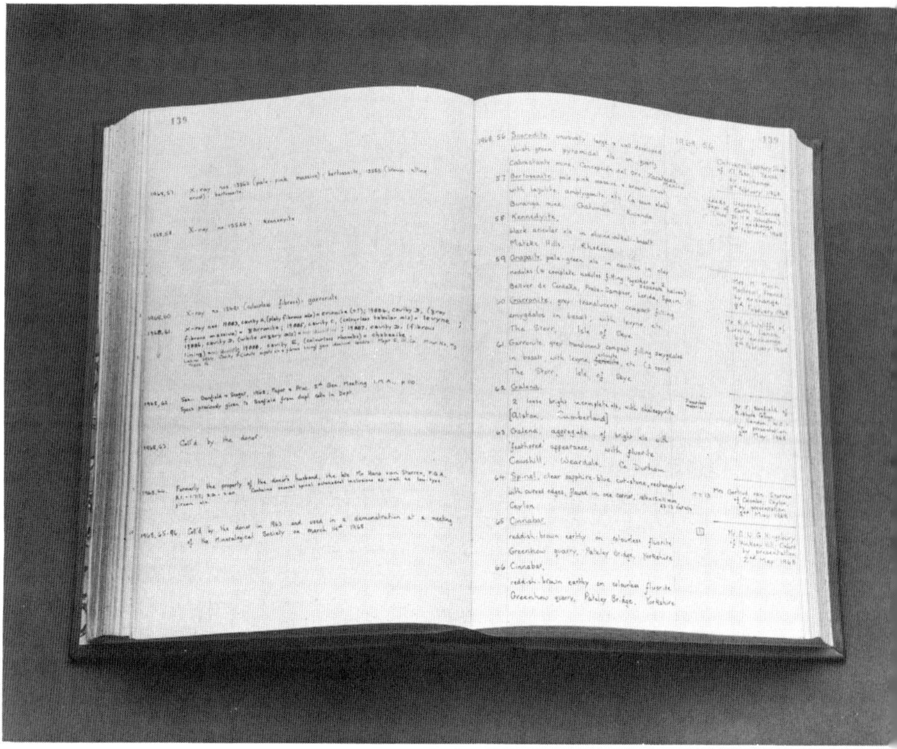

Fig. 9 A permanent catalogue of a collection should be kept recording details of the type of specimen, locality from which collected, date of collection or acquisition etc. The photograph shows a spread from one of the 'Mineral Registers' of the British Museum (Natural History). Details of specimen, locality, donor etc., are recorded on the right-hand page, and any data obtained from the specimen (analysis, X-ray etc) on the left

lubricant, because the label may easily become detached and lost. An alternative adhesive is polyvinyl acetate (p.v.a.) emulsion (Vinamul N9164) which, whilst water-based, dries to a water-resistant, slightly elastic solid. It is easily dissolved by acetone or cellulose thinners.

Another method of numbering specimens is to write the number with waterproof ink on a small area of enamel gloss paint (fig. 7). Further protection is provided by a coat of clear lacquer.

Probably the best method of storing specimens is to place them in individual, shallow cardboard trays, which are then stored in banks of drawers (fig. 11). The trays serve the double

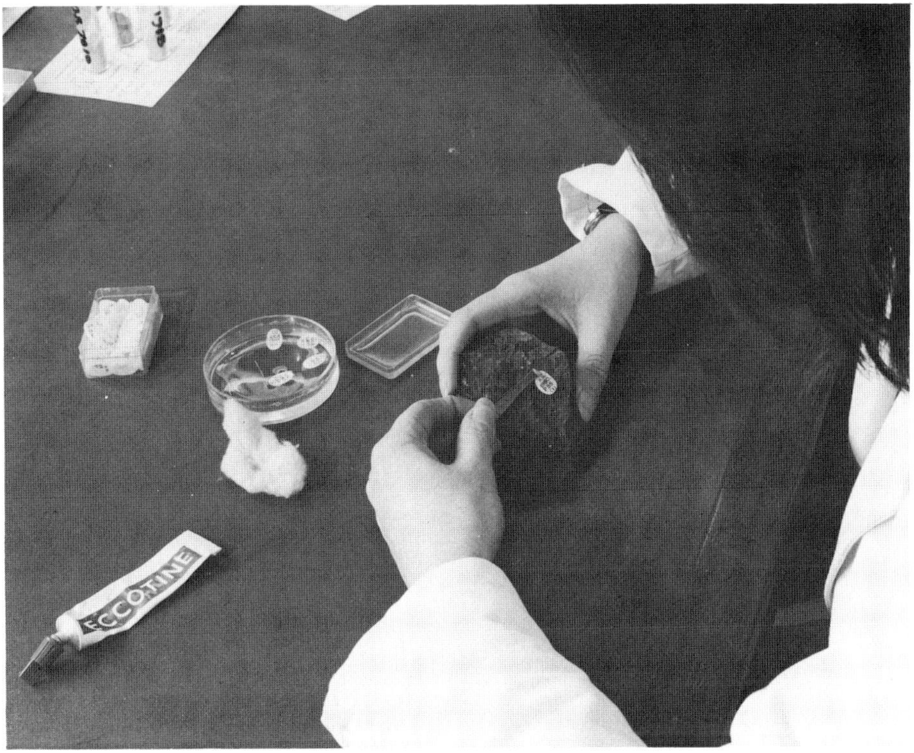

Fig. 10 Attaching a number ticket to a rock specimen. The ticket has been softened in water so as to mould more closely to the uneven surface. A little cotton wool mixed with the 'Seccotine' placed beneath the ticket strengthens the bond

function of keeping together specimen and labels, and protecting the specimens by preventing them rolling about and banging together when drawers are opened and closed. Drawers that contain only a few specimens should be filled with empty trays to prevent them sliding about. To protect fragile specimens it may be necessary to line the tray with some soft material, which should be 'lint-free', i.e. non-fibrous. Acid-free tissue paper is ideal, but cotton wool should not be used unless kept from contact with the specimen by tissue paper or some other lint-free material.

There are many ways in which collections can be arranged, but for minerals perhaps the most universally adopted is to keep minerals of one species together and divide these according to locality; this is the method adopted in the British

Fig. 11 A drawer of mineral specimens. Note the empty trays to prevent rolling and sliding when the drawer is opened and closed

Museum (Natural History). Rocks may be arranged according to type or kept in suites from particular localities or geological units.

Fossils

Collecting

Recording

Map, camera, notebook, sketch-book (fig. 6).

It is essential that accurate and detailed records should be kept of all specimens collected, and their localities accurately marked on a large-scale map. The exact position in the section from which the specimen was obtained should be recorded with a sketch, if necessary. Date each entry and record the map reference. Every specimen must be numbered to correspond with its notebook record. Scratch the number on the matrix (if practicable) and write a label with a waterproof, non-fading medium, e.g. waterproof ink, indelible pencil, 'Permanent-Black' felt tipped marker (fig. 7). The label itself should be water resistant. Many collectors mark adhesive tape and stick this to the matrix, but not to the specimen itself, since it may be damaged when the tape is removed. Wrap the label with the specimen. If put into a collecting-bag, the number should be added to the locality-label attached to the bag. Never rely on marking only the wrapping.

Stratigraphy and palaeoecology

When a section exposes more than one fossiliferous band, keep the fossils from each stratum apart and note their exact sequence. Make a sketch or diagram and photograph the section, measure the thickness of the various beds, number them and include the number on the labels of the associated specimens. Use a numbered bag for each horizon if collecting small specimens. Distinguish between fossils from the regularly bedded rocks of a section and those from infilled fissures. Limestones in particular should be examined for such fissure-deposits which can readily be distinguished by the irregular, chaotic arrangement of their infilling.

Palaeoecological records should be made including the distribution, association, preservation, relation to the sediment; the relative abundance and orientation of the fossils, and the top of the specimen as well as an arrow to indicate 'true' north. Follow the precept – eyes first, hammer last!

Extraction

Pretreatment

As a general rule preparation in the field should be minimal since it can be better done in the laboratory or at the home workbench; clumsy pre-treatment can make the final preparation difficult and time consuming. Nevertheless, many specimens are friable and/or badly cracked and need attention before extraction.

Having exposed as much of the fossil as possible, dry specimens should be treated with a thin solution (10 per cent wt/wt, or less) of (1) 'Butvar B98' in isopropyl alcohol; alternatively, (2) methylated spirit can replace the isopropyl alcohol, but 10 per cent volume/volume of diacetone alcohol must be added. Solution (2) is blue but any staining of the fossil will disappear in several days if exposed to light and air. The addition of diacetone alcohol prevents the formation of a white bloom in the dried plastic film (caused by the presence of water in the methylated spirit). Similarly, solution (1) with diacetone alcohol added can be used to treat slightly damp fossils without producing an obscuring bloom.

Expanded polystyrene dissolves readily in acetone or 'cellulose thinners' and makes a cheap, effective consolidant or 'hardener'. Solutions of shellac in methylated spirit, or cellulose acetate in 'cellulose thinners' have disadvantages but may be used effectively. Avoid the use of waxes and animal glues or sizes.

Solutions should be thin enough to penetrate the surface of the specimen, and thick surface films avoided. Several coats may be necessary, but allow each to dry before further application. When dry, the fossil can be extracted and if necessary, the other side may be 'hardened' before packing.

Large invertebrates and small articulated vertebrates can be treated by an adaptation of the techniques used for the larger vertebrates (see appropriate section) or by use of polyethylene glycol (p.e.g.) 4000, a water-soluble wax. This should be melted carefully (over a fire or camping stove) or carried molten in a Thermos flask and applied to the exposed and previously hardened surface of the fossil. On cooling it forms a rigid, closely adherent 'backing' giving considerable protection to fossils of up to 300 mm maximum size. A thickness of about 20 mm is adequate and it is best applied in several layers which cool more rapidly and reduce the effects of shrinkage. Incorporation of a few layers of cotton surgical gauze reduces the brittleness of the p.e.g. The removal of the backing is illustrated in Figure 18, p. 40.

Fossil and matrix

Fossils occur in various forms, as solid objects and as natural moulds. Casts from the latter can be prepared both in the field (fig. 12) and at home by using several materials. Damp clay or 'Plasticene' well worked to give a soft, cavity-free texture and pressed into the mould will give a good impression, although the 'squeeze' cannot be pulled entire from undercuts, and great care is needed in packing and transporting the results. Plaster of Paris poured into the natural mould gives a good result reasonably quickly but 'undercuts' are even more of a problem. Even the kitchen jelly-cube can be employed in an emergency if means are

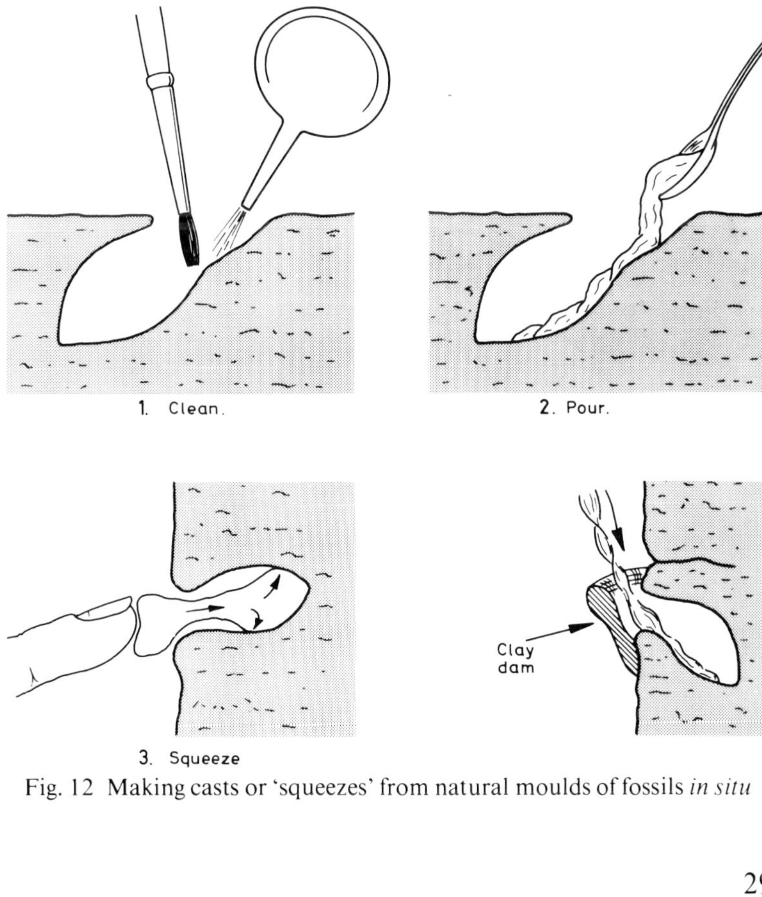

Fig. 12 Making casts or 'squeezes' from natural moulds of fossils *in situ*

available for melting, or carrying it molten. P.V.C. 'impression compounds' such as 'Fimo' are better than 'Plasticene' in that after removal from the mould, heating to approximately 150°C for 30–40 minutes produces a permanently hard product. A dusting of talcum-powder in the natural mould generally aids release of the impression compound.

Dental impression compounds, usually polysulphide (mercaptan) or silicone 'rubbers' are (apart from cost) ideal for smaller cavities. They set within a few minutes to give stiff rubbery products and the result is near-permanent. In very cold conditions warm the components before mixing. An added advantage is that water is used both as a separator and accelerator for the process. With adequate time any of the standard moulding materials – rubber latex, silicone rubber, etc. are suitable. Fossils are often difficult to distinguish in compact limestones, being more apparent on weathered surfaces: the calcite or aragonite of which many invertebrate fossils are composed may weather less rapidly than the limestone matrix. The apatite of fossil bone and the silica by which fossils of all kinds are sometimes replaced, are even more resistant. Old quarry faces, natural sections and stone-walls should therefore be examined with care. Examine apparently unfossiliferous limestones etc., with a hand lens; minute organisms such as foraminifera and Radiolaria, sponge-spicules and conodonts can often be seen, particularly if a smooth section of the rock is wetted. In most slaty rocks the planes of cleavage do not coincide with the original bedding-planes, which may sometimes be detected by colour-banding. Fossils must be sought on planes parallel with these and not necessarily the cleavage planes.

Nodules and concretions are worthy of special attention (fig. 13), since they often form around organic remains, and the best-preserved fossils or indeed the only fossils may occur in them. Clay-ironstone nodules associated with seams of coal may contain well-preserved remains of coal plants, insects and crustaceans. Nodules from the Gault (e.g. at Folkestone), cementstone nodules from the Lias (e.g. at Charmouth) and the Old Red Sandstone (NE Scotland) are often fossiliferous. Hollows in chert nodules and flints may contain a powder which includes sponge spicules and/or foraminifera; while the surfaces of nodules, pebbles and shells may bear

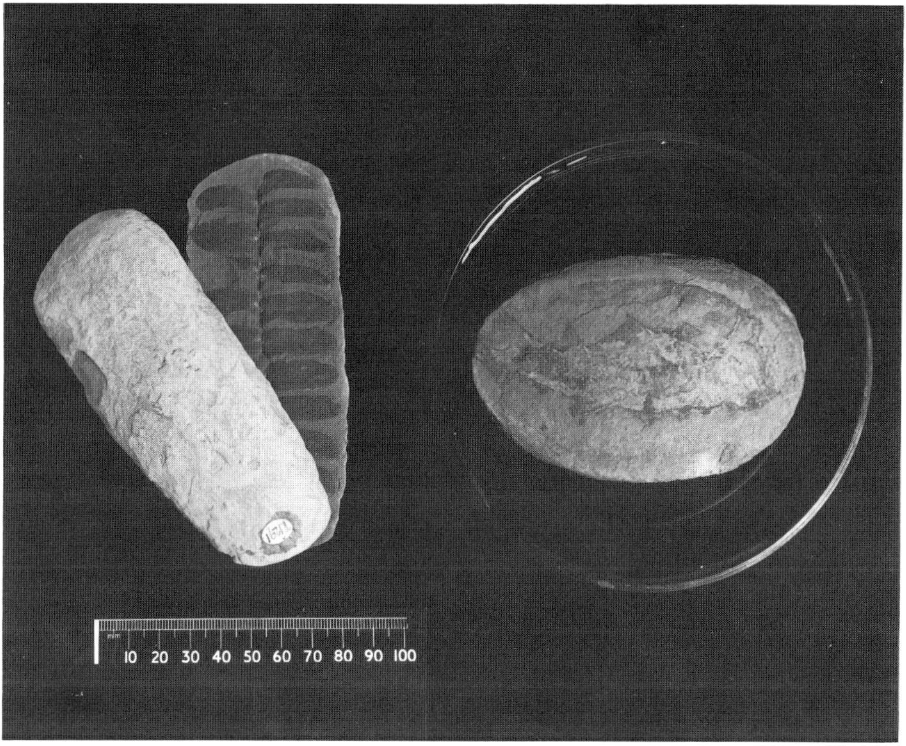

Fig. 13 Fossil-bearing nodules: on the left a plant, *Neuropteris gigantea,* Sternberg from the Coal Measures at Coseley, South Staffordshire, and on the right, immersed in liquid to improve definition, a fish, *Cheiracanthus murchisoni* Ag. from the Middle Old Red Sandstone, Lethen Bar, Nairnshire

encrusting organisms (e.g. polyzoa) that can also sometimes be found by brushing soft chalky limestone under water.

Sampling

Methods of sampling for micro-fossils are described in specialized text-books, but spot sampling at regular intervals (per 1500 mm) may be employed for thick sections of uniform lithology; intercalated clay or shale bands necessitating additional spot sampling.

Large foraminifera, conodonts and ostracods may be observed on bedding planes with the aid of a hand lens, but small ones, together with diatoms, radiolaria, spores and pollen, can only be distinguished with the aid of a microscope.

Great care must be taken to avoid contamination of samples, and sections should be brushed and scraped with clean tools before taking samples of (usually) 200 g, or about 0·5 litre in volume. High concentrations of fossils may call for bulk sampling for later preparation and analysis. The orientation of samples of indurated blocks should be indicated in addition to normal labelling.

Limestones may be collected in bulk for solution in acid (see Preparation, p. 44) to provide concentrations of bones, teeth and silicified invertebrates.

Larger vertebrates (fig. 14)

Although complete skeletons of the larger vertebrates are rare, isolated bones may frequently be found on the beaches below the Lias cliffs of Charmouth, Dorset or Whitby, Yorkshire.

Larger, stronger bones occurring *in situ* in soft sediments may be excavated carefully, lifted, wrapped with a generous quantity of newspaper and tightly packed in crates or deep trays, and surrounded by padding material such as wood-wool, straw, or plastic foam. Compression of the sediment over a long period frequently causes fracturing of the bone despite its apparent sturdiness. Moreover, when uncovered, the fossils are normally damp, and less 'mineralized' bones are likely to warp and crack badly on exposure to the air. In addition many specimens are badly leached and therefore soft or friable, especially those from peat, sand or gravel deposits, and these require special attention, particularly when laboratory preparation is likely to be delayed. Articulated skeletons must be protected against rapid desiccation and physical damage by covering with plastic-sheeting during excavation and the position of individual bones carefully recorded before being lifted singly, or in blocks.

In soft sediments the following technique should be adopted (fig. 15). The bone (or bones) is left standing on a pillar of matrix about 300 mm in height. Place two layers of tissue paper over the bone, dampen with water and with a soft brush mould gently to shape of the bone. This tissue prevents the next layer (plaster of Paris bandage) adhering to the bone, yet enables the plaster to conform to its profile. Undercuts and deep cavities are filled with wet, pulped newspaper, gently pressed in to even-up the profile and so avoid the formation of

Fig. 14 A plesiosaur, *Cryptocleidus,* found in the Oxford Clay at Fletton, near Peterborough, Northamptonshire

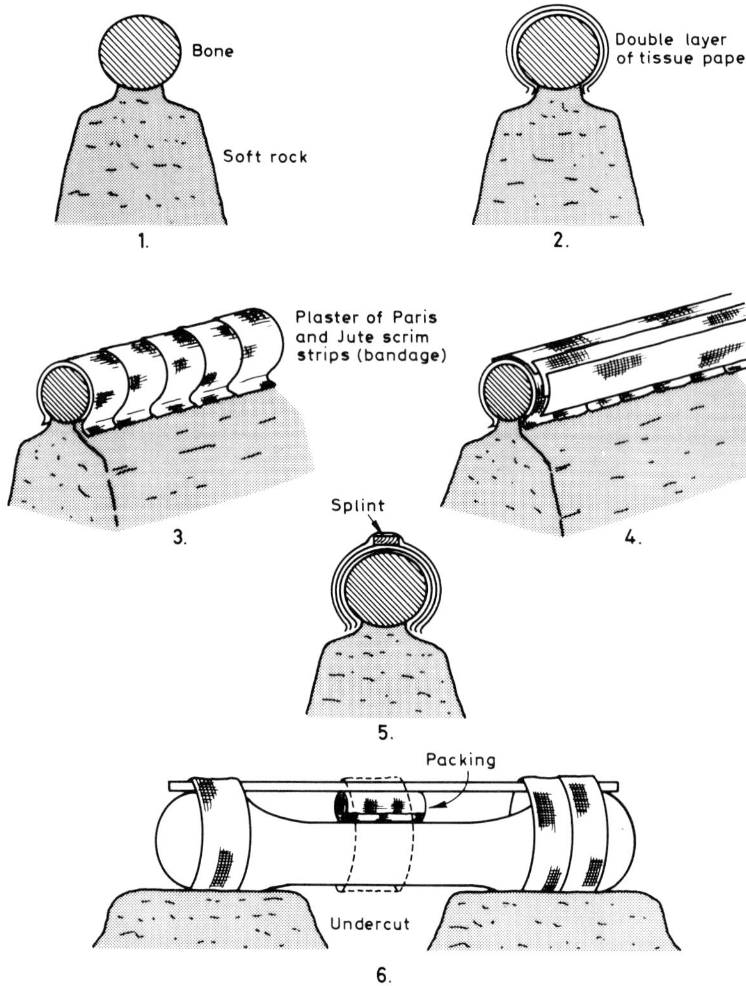

Fig. 15 Six steps in cocooning a fossil bone in plaster of Paris bandages

a plug of solid plaster of Paris, which would be extremely difficult to remove later without damaging the specimen.

Strips of plasterers' scrim (open-weave jute scrim or burlap) 90 mm wide are dipped into freshly mixed plaster of Paris and drawn out between the fingers to remove surplus and applied to the length and breadth of the tissued bone. The strips should not be longer than 500 mm (longer strips are difficult to handle and may make later removal awkward). Overlap the strips and build up two to three layers for bones up to 1·5 m in length. At this stage rigid support is supplied in the form of 250 mm section wood splints. (The dimensions, and material of the splints are obviously chosen according to circumstances, but the information given is appropriate to the size of fossil described.)

Orientation of the splints is again dictated by common sense, but as a guide, one splint along the centre-top is sufficient for bones up to 90 mm by 900 mm; two splints, one on each side for longer specimens, and further cross-pieces for wider and thinner bones. The splints are initially held in place by short strips of plaster scrim. Gaps between splint and plastered strips are filled at intervals with rolls of plaster scrim to prevent flexing during transportation.

Dig carefully through the matrix pedestal, in the centre beneath the bone. If the bone is well preserved the matrix can be removed to reduce weight, otherwise sufficient matrix should be left to provide additional support. Next apply the double layer of damp tissue to the newly exposed bone or matrix, overlapping on to the earlier plaster scrim. Further plaster scrim is then applied and wrapped around the specimen and splint(s), again in two to three layers. Fill up the tunnel in the matrix pedestal with debris to support the bone (but do not disturb the fresh bandaging) and repeat the process, first at one end of the pedestal then the other (with intermediate positions if required) until the finished cocoon rests on a pile of debris (fig. 16).

The completed cocoon must be left to dry before removal, a few hours on a fine summer's day usually being adequate. However, an excessively warm, dry atmosphere may result in the loss of the moisture needed for crystallization and 'set' of the plaster, and in such conditions the cocoon should be sprinkled periodically with water and covered during the initial stages of the setting process.

Fine industrial plaster of Paris sets quickly (i.e. 5–10 minutes) therefore only small quantities sufficient for immediate use should be prepared. Alternatively, one teaspoonful (5 g) of citric acid crystals per 4·5 litres (gallon) of plaster-mix, stirred into the water before adding the plaster of Paris, will retard 'final-set' by about an hour. The same amount of citric acid dissolved in a little water and mixed into plaster which has reached 'first-set' (point of sudden increase in viscosity) extends the viability of the mix. Conversely, an equal quantity

Fig. 16 Bones of the straight-tusked elephant found in Pleistocene clay at Aveley, Essex, are shown in various stages of plaster-bandage cocooning

of zinc sulphate accelerates the setting, and a 10 per cent solution in water sprayed or sprinkled on to badly set plaster (i.e. soft and crumbly, owing to rapid loss of water before complete crystallization) will harden it appreciably.

Ammonia solution (0·88 sp. gr. ammonium hydroxide, 5 per cent in water) added at the rate of 40 ml per 4·5 l (about two tablespoons per gallon) to plaster-mix retarded by citric acid, will quickly restore normality. The important point to remember in plaster-cocooning is that the purpose of the plaster-bandage is to contain the fossil, to protect it against further damage and to allow it to dry out slowly. Strength and rigidity are provided by the splints, so on no account build up a thick plaster-coat that is virtually impossible to remove without damaging the specimen. If splints are not available, make the cocoon as already described, but over the last layer of plaster-bandage place a 5 mm layer of wet newspaper. On

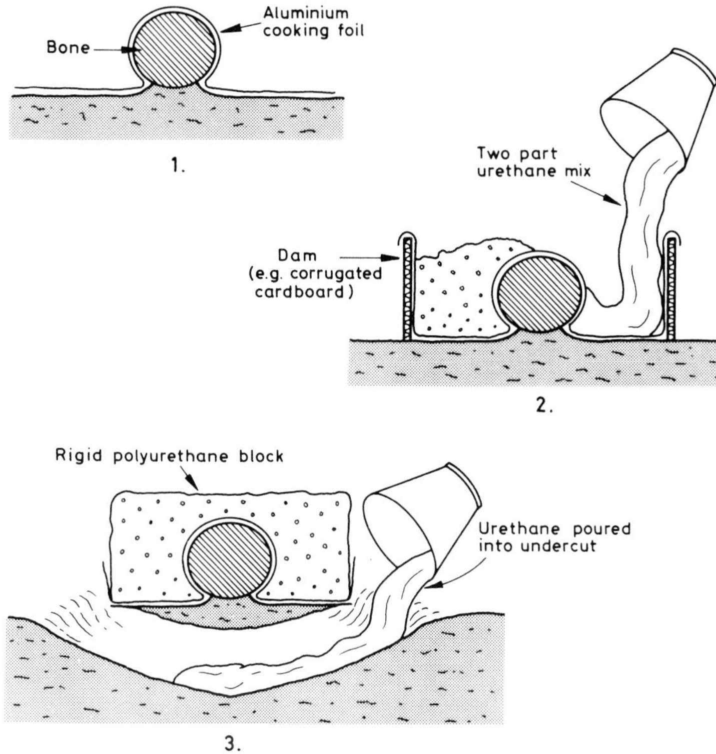

Fig. 17 Three steps in cocooning a fossil bone in rigid polyurethane foam plastic

this, 'first-set' plaster of Paris can be liberally applied to provide further support, the thick newspaper layer protecting the inner cocoon when the outer is chiselled away.

The removal of the specimen from the cocoon is illustrated in Figure 19, p. 41. Plaster-bandage cocooning is cheap and becomes an easy technique with practice. It is better for a 'dig' which exposes many bones, partly because of economy and partly because of the time needed for cocoons to dry. By the time the last specimen is fully cocooned the first has dried and is ready for removal.

When speed is imperative (and expense no object) polyurethane foam can be used (fig. 17). Excavate the bone as before, but without the pillar. Take a sheet of aluminium cooking foil and mould carefully around the exposed bone. Seal any tears with 'Sellotape' or masking tape, since polyurethane sticks strongly and, being virtually insoluble, is difficult to remove from the fossils. Build a dam around the bone (cardboard, wood, plastic sheet, metal or even piled up rock debris can be used), leaving 50–100 mm clear around the specimen to give a reasonable thickness of foam when poured. To avoid the possibility of the rising foam lifting the aluminium off the bone, the foil can be extended across the floor and up the inside of the dam-wall. Pour the urethane (mixed according to the manufacturer's instructions) into the bottom of the 'reservoir' and over the foil-covered fossil. The foaming mixture will rise to cover the specimen within 3–5 minutes and will set in 30 to 60 minutes.

Excavate beneath the specimen, as before, but ensure that there is a slight gradient down from both sides to the centre-line beneath the fossil (this is to direct the flow of foam-mix). Apply the foil and pour the foam-mix. Usually a dam is not needed at this stage, but by placing two pieces of card against the tunnel openings, supporting them against the pressure, and leaving a gap at the top, surplus foam can be directed upwards to overlap the top half of the cocoon. Splints are not necessary, setting is rapid, and the foam gives good protection during transport; however, wet specimens will not dry out through the cocoon as with plaster, hence long-term storage is not advisable.

Warning. Toxic fumes (one component is an isocyanate) are given off particularly during the exothermic foaming stage. Good ventilation is essential and this applies also to the transporting vehicle afterwards, because of the residual isocyanate in the foam.

Occasionally a fragile bone is found in a position which makes total excavation impossible. In this instance the exposed upper surface of the bone should be cleaned as thoroughly as possible, and hardened with the Butvar B98 solution (as described on p. 28), or polyvinyl alcohol (5 per cent w/v in water). When the coating is dry, a viscous solution of Butvar B76 (20 per cent w/v in acetone) is brushed on, followed immediately by a jacket of coarse glass-fibre cloth (e.g. 450 g (16 oz) woven-rovings), which is applied closely to the bone, using a brush and more Butvar solution. Fine cloth or matting can be used for delicate or difficult edges. Cylindrical bones usually need no further support, but small splints can be incorporated or hollow glass-fibre and resin ribs formed, to increase rigidity in flatter objects. The matrix that can be reached beneath the specimen is then excavated and the plaster or polyurethane cocoons completed as far as possible. When the fossil is lifted, the uncocooned areas remain cemented to the glass-fibre cloth, and the cocoon is completed. Since Butvar B98 and polyvinyl alcohol are not soluble in acetone, when the glass-fibre resin jacket is softened with this solvent in the work-room, the bone surface remains relatively unaffected and the glass-fibre can be safely removed.

In hard rocks, the bone exposed on the surface is protected as described above, but the enclosing rock is relied upon to protect the unexposed surface. Further preparation is generally impossible or inadvisable in the field, although where possible rock is chiselled away to reduce weight. Breaking out and transporting the blocks requires effort and ingenuity as well as great care for personal safety, and certainly in the case of large specimens this work should not be attempted without experienced persons being present.

Preparation (Development)

Mechanical

Many fossils require little more preparation than thorough cleaning (but see figs 18 and 19). Suitable specimens can be brush-cleaned using soft paint-brushes, tooth-brushes, typewriter-brushes, even scrubbing-brushes; dry, with water, or with water and detergent.

Needles are invaluable for removing less-indurated matrices, particularly from fine ornament. They can be of a wide variety of sizes, from fine entomological pins, or steel sewing-needles to headless, steel, cork-tile pins of 15 S.W.G. (1·8 mm), which are most conveniently held in engineers pin-vices or pin-tongs which are available in most ironmongers. A variety of dental hand instruments are of use, including chisels and cutters, and those tipped with tungsten carbide are particularly suitable for geological preparation.

Hammers and a variety of chisels are necessary for coarse and fine work. When working close to a fossil a 1·5 mm (3/16 in) chisel or preferably a 'pin-chisel' is used (a solid-shafted pin-vice and appropriate pin makes an excellent chisel) with a 'club' or 'lump'-hammer of about 225 g (0·5 lb).

The specimen must be placed on a shock-absorbent pad

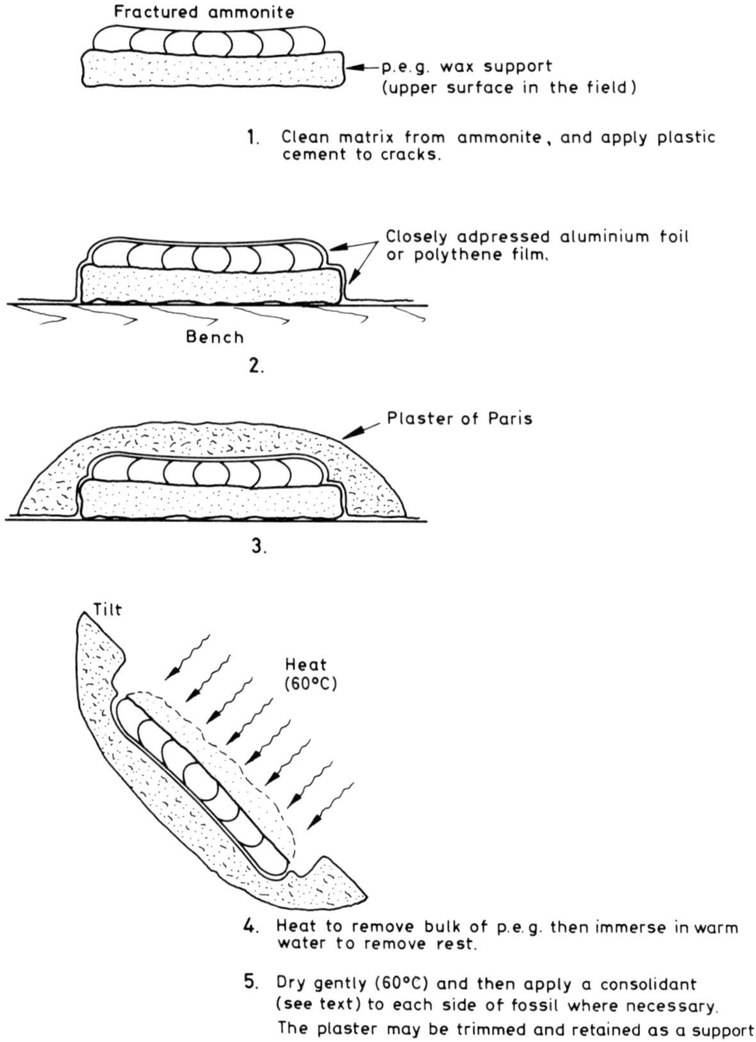

Fig. 18 The five steps to be followed in removing a polyethylene glycol (p.e.g.) wax field support

such as foam-plastic or a loosely filled sandbag which prevents the fossil shattering under the impact of hammer and chisel and also prevents it from slipping. The chisel is held at an angle of about 70 degrees, and the point placed on the matrix 1 or 2 mm away from the exposed surface of the fossil. The hand holding the chisel rests on the fossil or its matrix and is tensed so that after the butt of the chisel has been struck

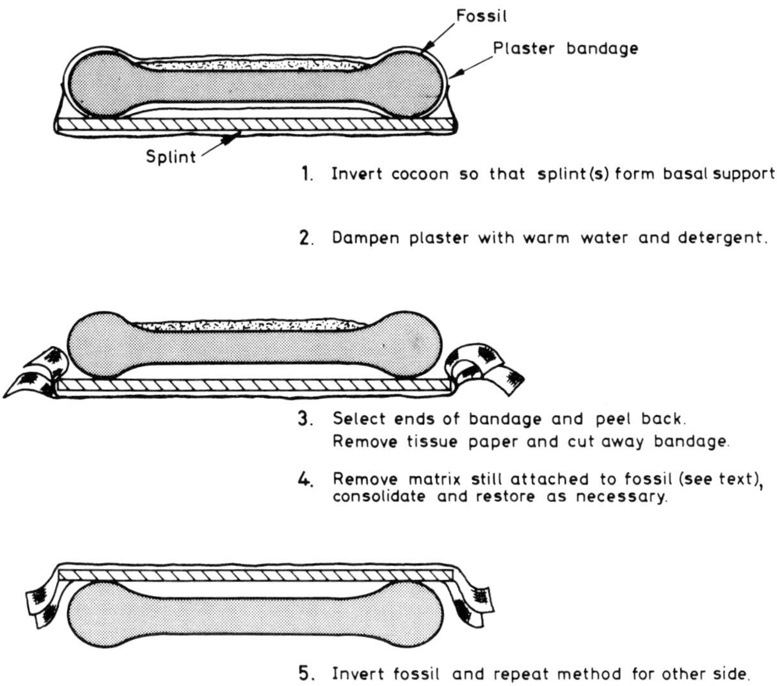

Fig. 19 The five steps to be followed in removing a plaster of Paris bandage field cocoon

by the hammer and a flake of rock detached, the chisel point rebounds and damage to the fossil is avoided. The chisel is moved sideways and the process repeated until the fossil has been traversed. Then the chisel is moved back a little and the action repeated until the fossil has been completely cleared.

The more sophisticated and expensive equipment used by professional preparators is worthy of mention. Many keen amateurs can afford at least some of the equipment, some of which can be obtained second-hand.

Several mechanical chisels are available in a wide range of size and power. A simple and reasonably cheap electric engraving-tool is available in most ironmongers or DIY shops, with an adjustable stroke and a tungsten carbide engraving tip. A pin-chuck adaptor is also available. Electric drills may be adapted to percussion drilling, as well as operating flexible drives with chucks to hold abrasive burrs for coarse preparation.

A dental workshop motor with a flexible drive-shaft, foot-control, a choice of hand-pieces and a selection of burrs and brushes is most useful. High-speed rotary hand-pieces with ball-bearings are preferable to those with plain bearings, although they are bulkier and more expensive. A slip-chuck, straight hand-piece (No. 2. Slipjoint) is desirable, enabling rapid change of hand-pieces and burrs.

A wide variety of burrs, mounted-points, discs and cutting wheels is available, in steel, tungsten carbide, and diamond. Steel burrs are suitable for very soft rocks only, as they wear rapidly; while diamond burrs demand a high-speed motor. Rotary hand-pieces can be used for holding dental brushes of various shapes, materials and grades of hardness; some with fixed mandrels, others requiring a suitable mandrel adaptor (e.g. Porte-polisher). A stiff bristle brush used at high speed will remove a wide variety of matrices, including friable sandstones, dry clays, and especially chalk.

An adjustable dental mallet, preferably with a pin-chuck adaptor, can also be used with the flexible drive, providing a more comfortable and sensitive instrument than an engraving-tool.

Although fine and delicate work is possible with the chisel tools, particularly when used in conjunction with a shock-absorbent pad, some fossils e.g. teeth in a small mammalian jaw, require temporary support at some point if all surfaces are to be cleared. Polyethylene glycol 4000 is an ideal support in most cases; it is a white, solid wax melting at 55°C and is soluble in water. Hence no pressure is required to apply the support, since the wax is melted and brushed on or poured, and it is removed by immersion in water.

Modern air-driven dental drills combine very high cutting speed (500–500 000 rpm) with little torque, making this very safe for both operator and diamond tools. The piston-driven 'high torque' air drills give a range of speeds (100–60 000 rpm) with sufficient torque to drive cutting and grinding discs.

Several manufacturers now supply pneumatic engravers of the size of fountain pens employing compressed air ($2{\cdot}8\ kg/cm^2$–$7{\cdot}0\ kg/cm^2$), which provide a superior alternative to the electric types already mentioned. They are more comfortable to use, more power is available and the exhaust is directed over the point, cleaning debris away as it forms. Again, a pin-chuck increases the pen's versatility. Control is by varying the volume and/or air pressure, and some makes are provided with a valve on the pen. Bigger pneumatic and electric hammers are available for the initial stages of preparation of large nodules.

A modern method of mechanical development is by means of ultrasonic probes. A dental device provides a series of insert-tips for various jobs, most using water, either sprayed along the tip or by total immersion, to produce 'cavitation' bubbles to clean the specimen. The most useful tip has proved to be the dry 'Root Canal–PR–30' which utilizes a small chuck to hold fine needles. The movement of the point is so small (0·025 mm) that very fine structures can be developed in conjunction with a binocular microscope. Although dental equipment is expensive when new, its second-hand value is low, and bargains are advertised in dental magazines which can be consulted in many public libraries.

Ultrasonic tanks are much used to remove clays, sands, soft limestones (chalk) etc., where unique specimens are not involved, the fossils being left to develop without close supervision. Large tanks are used for bulk cleaning, while individual fossils can be placed in beakers or bowls in the larger tank. Detergent added to the cleaning water helps and a variety of chemical solutions (see pp. 44–45) can be used for particular problems. If the agitation is too violent, the use of plastic-ware rather than glass containers reduces the effect. Chalk echinoderms, non-pyritic Lias Clay cephalopods, Oxford Clay reptile bones, etc. are effectively cleaned by this method. Ostracod valves and diatoms are separated effectively by short immersion – 10 seconds for the latter. Ultrasonic tools are invaluable but are a potential danger to hearing and should be provided with acoustic-cladding. Otherwise the operator must wear ear defenders and use the tools in an area isolated from other persons. Vibration-free cutting is possible with the 'Airbrasive', a device which delivers fine abrasive particles (10–75 micron) in a stream of compressed air (or gas). The gas pressure, particle feed and hardness of the abrasive may be varied to suit various matrices. Exposed surfaces can be protected using wax or rubber coatings, but it is difficult to prevent rapid erosion of soft fossil material by abrasive passing through minute holes in the enclosing harder matrix – by the time the hole is noticed, the underlying fossil has gone!

Chalk echinoderms can be cleaned quickly and easily using

sodium bicarbonate or dolomite abrasives. Although resistant matrices can be removed from vertebrate fossils using harder abrasive powders, the 'Airbrasive' is best utilized as an adjunct to the acid-preparation technique (see below) to remove softened or insoluble residues from the surface of the fossils. Grinding tools and sand blasters (inc. the 'Airbrasive') produce great quantities of dust. In professional laboratories, efficient air extraction and filters used in conjunction with glove-boxes and fume-cupboards are obligatory. At home the use of an efficient filter face-mask is the alternative. Always wear goggles to protect the eyes when chipping and grinding rock.

Chemical

Acids are used in two ways – either to remove the matrix and free the fossil; or to dissolve the fossil and leave a natural mould in the matrix, from which a cast may be prepared. Thus phosphatic fossils (inarticulate brachiopods, conodonts, bone) can be extracted from limestones with acetic or formic acids (2–15 per cent solutions); whereas carbonate and phosphatic fossils in sandstone with a siliceous cement (e.g. some Old Red Sandstone rocks) are dissolved in hydrochloric acid (10 per cent), leaving the natural mould.

Carbonaceous (plant) material, chitin, and specimens preserved in pyrite can usually be freed from limestones by acetic acid. Complexing or chelating agents (versenes) such as EDTA or its salts can also be used to remove carbonates. The use of a buffered 5 per cent aqueous solution of thioglycollic acid will effectively leach away iron oxides to release fossil bone from ironstone nodules.

Warning. All the above acids are dangerous and must be used with the greatest caution. (See 'Safety precautions'. p. 50).

Salt solutions are occasionally used to break up porous, friable rocks by the disruptive effect of crystal growth. The rock is soaked in a warm, supersaturated solution of sodium thiosulphate (hypo), sodium acetate, or magnesium sulphate, and when the solution is cool, a crystal of the salt is dropped into it causing rapid crystallization. The growing crystals fracture the porous rock and the process is repeated until the rock has disintegrated. Although normally used for microfossil recovery, the technique can be applied to macrofossils.

An alternative method is alternately to freeze and thaw water in which the rock is immersed. Ironstone nodules can be split by heating and then plunging into cold water.

Detergents, dispersing and foaming agents and solvents, such as petrol (gasoline), paraffin (kerosine), acetone, etc. may be used to break down clays, shales and soft sands.

Hydrogen peroxide can be used to disintegrate soft matrices by the vigorous production of bubbles. A few drops of dilute (0·1 M) ammonium hydroxide, or 'household ammonia' in 20–40 vol. hydrogen peroxide accelerates the production of oxygen bubbles.

Conservation

Some fossils and rocks are subject to decay and require appropriate treatment. However, consolidants or other chemicals should not be applied indiscriminately since they may have to be removed later.

Preparation

Develop and prepare each fossil as thoroughly as possible. However, if the specimen is left in a block of matrix, the exposed surface should be as clean as possible to show all available detail, and also to prevent possible decay starting in the fragments of attached matrix (particularly pyritic matrices). Fossils from the foreshore may be saturated with sea-salt which will accelerate the destructive process of pyrite decay and possibly cause mechanical damage by crystal growth under humid conditions. This may be removed by washing and the specimens should be thoroughly dried. Note that clays and shales, peat, soft sands etc. disintegrate completely in water, and so cannot be treated in this way.

Consolidation

Friable, fragile fossils can be 'hardened' or consolidated, using solutions of resins, or resin emulsions. It is a rule of conservation to use a reversible process and the following are recommended:

Polyvinyl butyral resin solutions:

Butvar B90 (or B98) 1 pt : 9 pts (w/v) isopropyl alcohol

Butvar B90 (or B98) 1 pt : 8 pts methylated spirit : 1 pt diacetone alcohol.
(for acid preparation use polybutyl methacrylate (Vinalak 5911) or Bedacryl 122X, both diluted 1 pt : 4pts methyl ethyl ketone).
These work well on dry fossils, and the addition of 1 per cent diacetone alcohol to the first solution will enable the treatment of slightly damp specimens, avoiding an opaque, white discoloration. As alternatives, polyvinyl acetate can also be dissolved in the alcohols mentioned above, or less desirably, shellac or brillac.
Polyvinyl acetate emulsions:
Vinamul N9146 1 pt : 3 pts water
This is ideal for damp specimens, fossil wood (non-pyritized) and ivory. The emulsion may be further thinned to aid penetration. Ivory and wood treated this way should be dried slowly; a tent of polythene over (but not touching) the fossil will ensure this. Repeat the process as necessary; aim to fill all cavities and pores but avoid a thick, obscuring coat on the surface, removing pools and tears of p.v.a. after each coating. Large cracks may be filled with neat emulsion. The emulsion can be applied by brushing, spraying or dipping, but immersion requires caution especially of poorly preserved fossils or specimens that have been previously consolidated with water soluble glues.

Drying slowly in the way described, generally prevents warping and cracking. Allow 1–3 hours between applications, and at least 24 hours after the final coat remove the tent and dry at room temperature for a further 2 or 3 days. Then, if required, use a Butvar solution as previously described. Long bones, and ivory in particular, may require the use of tourniquets to reform and hold warped segments during the process of 'hardening'.

Repair

Glues or cements for repair and gap filling are numerous but should be easily soluble in a convenient and safe solvent. 'Butvars' make a quick-drying cement in ethyl acetate or dichloromethane, and the 'B76' is soluble in acetone. Polymethyl methacrylate, e.g. Perspex, Lucite and Plexiglas, dissolves in dichloromethane or ethyl acetate, and is an invaluable quick-drying cement for small bones especially in acid preparations. Polystyrene, easily obtainable as scrap in its expanded form (ceiling-tiles, etc.), dissolves almost instantaneously in acetone, dichloromethane and ethyl acetate, and is a good cementing medium.

Proprietary glues such as 'UHU' (polyvinyl acetate) and Lepage's polystyrene and balsa (cellulose acetate) cements are useful and also conveniently packaged for occasional use. P.V.A. emulsion, already mentioned, is an excellent adhesive for smaller specimens and can also be obtained easily in small proprietary packs, e.g. Evostik W.

Large cracks can be filled with Fibrenyle or A.J.K. dough; both are resin/water emulsions filled with absorbent fine fibres and, in A.J.K., kaolin, to give a ceramic finish. Like all resin/solvent mixtures there is some shrinkage on drying and so this should be applied in layers, allowing each to dry before adding another. Industrial methylated spirit (i.m.s.) or iso-propanol can be used as a lubricant when working the dough into a smooth finish or as a solvent for the resin component. Ethyl acetate or dichloromethane are rapidly acting solvents which, added liberally to the fresh dough, make a fast-drying and very strong cement, particularly useful for repairing large, heavily-mineralized dinosaur bones.

Plaster of Paris is still considered an excellent filler being cheap and easy to use, though with the disadvantages of introducing a great deal of water to the specimen, providing only mechanical adhesion, and requiring removal by mechanical means. However, adhesion can be improved by the addition of 1 per cent (v/v) p.v.a. emulsion to the plaster-mix and at the same time painting the cavity to be filled, with neat p.v.a. emulsion. Soaking set plaster of Paris with a 10 per cent solution of sodium citrate in water (hot if possible) will soften and crumble the plaster if it needs to be removed, but the fossil will need to be thoroughly washed, soaking in water to remove any absorbed salt. 'Polyfilla' is an alternative to plaster of Paris with the added advantage that it does not require prior sizing or wetting of the material to be filled. Mixed and kneaded as a stiff dough, there is little water in the Polyfilla and shrinkage is thus reduced. Removal is also easier than with plaster of Paris as it softens in water.

A temporary filler which can be used as a support during preparation or support/sealer during moulding, is poly-ethylene glycol 4000. It can be melted and poured, or used in a 'Plasticene'-like form. Both are removed easily in water. Cavity filling in acid preparations requires a cold-setting acrylic material (2-part), such as N.H.P. modelling plastic, the liquid being added to the powder in proportions giving the required consistency – more liquid for cracks, more pasty for holes. However, shrinkage occurs over a long period, and the addition of 100 per cent (v/v) of glass micro-spheres or silica powder is recommended to reduce this. Take precautions against breathing in these dusts.

Pyrite conservation

Fossils wholly or partially preserved in pyrite or set in pyritized matrices require special attention. Recent research suggests that the oxidation and breakdown of pyrite is due to chemical processes not to bacterial action. The chief oxidation products are sulphuric acid and various hydrated sulphates, mainly of iron. These sulphates are usually visible as efflorescent growths on the surface and can cause total disruption by growing rapidly in the interstices of the affected material. Furthermore, the acid destroys bone or shell and the labels, boxes and wooden drawers of the storage cabinets. High humidity (60–75 per cent r.h.) is the major factor in initiating pyrite decay and must be avoided. If decay has started remove affected material immediately to a dry environment. Neutralize acid by exposure to ammonia in a closed vessel, the usual source of ammonia being a strong solution of ammonium hydroxide. Alternatively, a 5 per cent solution of morpholine in isopropanol or industrial methylated spirit may be applied, the process being repeated several times over a period of 48 hours. Ammonia and morpholine are toxic! Use in a well-ventilated area or wear an approved respirator.

Freshly collected fossils should be cleared of unwanted matrix and those from coastal sites that may contain salt should be soaked in several changes of distilled or de-ionized water and then thoroughly dried. Consolidants should be applied only when the fossil material is friable and then it must be one that is easily removable e.g. a Butvar/isopropanol solution. Consolidants cannot be expected to prevent pyrite decay!

Curation

The basic rules in packaging fossils for transport are: firstly, that any container must be packed to prevent internal movement; secondly that the packing material must be resilient to absorb shock and reduce compaction; thirdly, large, heavy cases should be avoided if possible since they invariably suffer more damage than smaller, lighter ones – and you may have to lift them yourself!

Newspaper is an adequate packing medium for most specimens but must be used generously and wrapped firmly before packing tightly into boxes. Smaller, finely ornamented fossils will benefit from a wrapping of tissue-paper (preferably unglazed and acid-free) before the newspaper. Never wrap fragile or highly ornamented specimens directly in cotton wool or other fibrous packing; it is difficult to remove and may well damage the surface during removal. Wrap each specimen or fragment separately if possible. Fill any spaces left in the boxes with crumpled newspaper, straw, moss, grass, wood-wool, cotton wool, or expanded polystyrene 'worms' or pieces. Never use sawdust, sand, earth or chaff, i.e. anything that is abrasive, packs down or can pour out through cracks in the case. When collecting outside Great Britain it is as well to check the regulations of the countries concerned before choosing packaging materials; some natural materials are prohibited because of the threat of the introduction of disease.

Plaster-bandaged field cocoons, if not packed in crates for local transport, require packing material between them and the floor of the transporting vehicle; while smaller cocoons need lateral packing and possibly a covering tarpaulin to prevent them bouncing. The expanded polyurethane cocoons are less vulnerable but require close packing nevertheless. Small, delicate fossils, e.g. vertebrate skulls which have been acid-developed, are easily crushed and are better embedded in polyethylene glycol wax of 4000 or 6000 grade (solidifying at 56°C and 63°C respectively). Fill individual cavities first and allow the molten wax to freeze, before applying brush-coats of the wax all over, to encapsulate the specimen. The specimen is then wrapped in tissue-paper and newspaper as before.

Storage of the collections is a problem for large museums and amateurs alike. The first requirement is that specimens should be readily accessible, accompanied by sufficient data to refer quickly and accurately to the full record in notebooks or registers. Specimen number, locality and horizon can be abbreviated and put on to the fossil itself, with labels and notebooks providing fuller record, i.e. '25.LC.4.H' on the fossil might indicate 'Specimen 25, London Clay, Division 4, Hanwell (Middlesex)'.

Well-fitted drawers in a chest or cupboard provide ready access to specimens contained in open trays, (with data label in the base) and protect the fossils from dust. However, the trays must be packed to avoid movement if the drawer is opened or closed violently by mistake. Transparent lids to the trays will prevent the fossils jumping out if the drawer is moved abruptly, and enable immediate inspection. Collections on open shelves can be stored with labels in transparent plastic bags to keep the fossils clean. A small amount of absorbent packing in the bag beneath specimens serves to protect them from knocks during movement and also guards against water condensation inside the bag. A desiccating agent e.g. silica gel, suitably conditioned, may also be enclosed, where appropriate, to prevent pyrite decay.

It is important to note that some hardwoods, e.g. oak, and glues, emit acid fumes in sufficient concentration to affect calcareous fossils. First signs of this attack are the growth of whiskery crystals over the surface of the fossil. This may be countered by increasing the ventilation to prevent the build-up of fumes, and including shallow trays of precipitated chalk to neutralize the fumes *may* help, but the only sure answer is to avoid such materials. Small fossils preserved as pyrite may be stored in stoppered jars immersed in silicone-fluid. Distilled water, paraffin (kerosine) and glycerol have proved to be ineffective. Silicone fluid (200/350 cSt.) is expensive initially but does not suffer from evaporation. It may, however, require the use of solvents e.g. dichloromethane or toluene to remove the fluid when the fossil is required for study (both solvents are harmful so use only in well-ventilated areas).

Chemical hazards

The use of chemicals in preparation and conservation requires some knowledge as well as common sense, and the supplier should be asked for guidance on handling the chemicals and the dangers involved. Large chemical manufacturers produce wallcharts of the commoner chemicals, the hazards involved and the methods for dealing with spillage and disposal.

Chemicals discussed in this publication are briefly commented upon in the table below.

Chemical	Flammability	Toxicity	Disposal Code
Acetone	Very high	Narcotic. Serious poisoning rare	1
Ammonia solution	Nil	Harmful liquid & vapour	2
Butvar, p.b.m. & other resins in solution	Dependent on solvent	See under solvent used	3a
Diacetone alcohol	Medium	Harmful liquid, low volatility reduces hazard	1
Dichloromethane	Nil	Harmful irritant in high concentrations. May cause cyanosis	3b
Ethyl acetate	Very high	Harmful liquid & vapour (prolonged inhalation)	3a
Hydrogen peroxide	Nil	Corrosive (oxidant, bleach) harmful liquid & vapour	4
Methyl ethyl ketone	Very high	Narcotic in high concentrations	1
Morpholine	Medium	Harmful liquid & vapour	1
Iso-propyl alcohol	High	Moderately harmful liquid and vapour	1
Toluene	Very high	Moderately harmful liquid and vapour	3a
Acetic acid (glacial)	Medium	Corrosive harmful liquid and vapour	1
Formic acid	Nil	Corrosive harmful liquid and vapour	5
Hydrochloric acid	Nil	Corrosive. Very harmful liquid and vapour	6

Disposal (Spillages)

1. Turn off sources of ignition. Wear gloves and face-shield or goggles. Mop up with plenty of water and run to waste, diluting greatly with running water. Ventilate area well.

2. As '1' ignoring sources of ignition.
3. (a) Turn off sources of ignition, (b) wear gloves and face-shield or goggles. Apply non-flammable dispersing agent (if available) and water, agitating with a brush to form an emulsion. Run to waste with copious supply of running water. Alternatively absorb into pile of sand, remove in bucket to open area and allow to evaporate. Ventilate area well. Clean site of spillage with detergent and water.
4. As '1', ignoring 'sources of ignition'.
5. Wear gloves, and face-shield or goggles. Spread soda-ash or sodium bicarbonate over spillage and mop with copious supply of water. Run to waste, diluting greatly with running water.
6. Wear breathing apparatus, otherwise as '5'.

It is assumed that in all instances small quantities of the above chemicals are involved, in which case the local authority may allow disposal after treatment, as indicated above, into the sewers, but check beforehand. HMSO publish many pamphlets, etc. on handling hazardous materials, including the *Deposit of Poisonous Waste Act 1972.*

Protective clothing must be worn when handling corrosive chemicals; rubber or plastic gloves and apron are essential and goggles or face-shield if the facility of the fume cupboard sash is not available. Footwear is also important if spillage is likely. Running water is essential to remove liquid from the skin and it must be possible to get the head under this source if eyes are affected, so if the source is over a basin or sink keep this as clear as possible. Do not store dirty glassware in it for later cleaning. Barrier creams give some protection to the hands, particularly against the degreasing effect caused by solvents, but this protection is less certain than that given by rubber gloves.

Further reading and sources of reference

Ager, D. V. 1963. *Principles of palaeoecology.* 372 pp. McGraw Hill Book Co. Inc: New York.

Brothwell, D. R. 1981 [3rd edn.]. *Digging up bones.* 216 pp. British Museum (Natural History): London.

Camp, C. L. & Hanna, G. D. 1937. *Methods in palaeontology.* 153 pp. University of California Press: Berkeley.

Himus, G. W. & Sweeting, G. S. 1951. *The elements of field geology.* 268 pp. University Tutorial Press: London.

Jones, D. J. 1956. *Introduction to microfossils.* 406 pp. Harper & Brothers: New York.

Kummel, B. & Raup, D. 1965 [editors]. *Handbook of paleontological techniques.* 852 pp. W. H. Freeman & Co.: San Francisco.

Rixon, A. E. 1976. *Fossil animal remains: their preparation and conservation.* 304 pp. Athlone Press: London.

Simpson, B. 1970 [Reprinted]. *Geological maps.* 98 pp. Pergamon Press: Oxford.

Materials, manufacturers and suppliers

Airbrasive (and accessories), S. S. White.
Supplier: G.E.C. Mechanical Handling Ltd.
A.J.K. Dough (and Fibrenyle), Ingredients.
Supplier: Joel, F. W.
Alginate moulding compounds (see Blueprint).
Supplier: Joel, F. W.; Claudius Ash & Sons Co. Ltd.
Anthogyr dental mallet no. 1650.
Supplier: Courtin, Marcel.
Bedacryl 122X. Manufacturer: Imperial Chemical Industries Ltd.
Supplier: Joel, F. W.
Blueprint. Manufacturer: De Trey.
Butvar, B76, B98. Manufacturer: Monsanto.
Supplier: Joel, F. W.
Carbowax 4000. Manufacturer: Union Carbide.
Supplier: Joel, F. W.
Cellulose thinners.
Supplier: Hardware, ironmongery and paint stores.
Chemicals (see Solvents and other chemicals).
Chisels, cold.
Supplier: Hardware and ironmongery stores; Tiranti, Alec. Ltd; E.L.E. Ltd.
Dental apparatus, tools and materials.
Supplier: Park Naval Engineering Ltd; Claudius Ash & Sons Co. Ltd; Cottrell & Co.; Courtin, Marcel; Morris, B. O. Ltd.
Engravers (see Vibro-pen and Vibro-tool)
Fibrenyle (see A.J.K.)
Fimo modelling compound.
Supplier: Joel, F. W.
Glass-fibre, all types.
Supplier: Strand Glass Co. Ltd; Tiranti, Alec. Ltd; Trylon Ltd.
Goggles, safety.
Supplier: Joel, F. W.; and some hardware/ironmongery stores.
Hammers, club and lump.
Supplier: Tiranti, Alec. Ltd; and some ironmongers.

Hammers, geological.
Supplier: Gregory Bottley & Lloyd; Geological Museum; lapidary (DIY) shops; E.L.E. Ltd.
Laboratory supplies, see: Gallenkamp Ltd; Griffin & George Ltd.
N.H.P. Mounting & Modelling Plastic.
Supplier: Howmedica.
Pins, steel, headless, 15 s.w.g. Manufacturer: Armstrong Cork Co.
Supplier: Specialist floor fitting shops.
Pin-vices, e.g. Eclipse range.
Supplier: Hardware and ironmongery stores.
Plaster of Paris, casting grades.
Supplier: Builders merchants; Strand Glass Co.; Tiranti, Alec. Ltd; Trylon Ltd.
Plasterers scrim (inc. Jute-scrim pipe-lagging in 4-inch wide rolls).
Supplier: (see Plaster of Paris).
Polybutyl methacrylate (see Bedacryl & Vinalak).
Polyester resins. Manufacturer: Scott Bader Ltd; I.C.I. Ltd.
Supplier: Strand Glass Co. Ltd; Tiranti, Alec. Ltd; Trylon Ltd; Joel, F. W.
Polyethylene glycol (see Carbowax).
Polymethyl methacrylate; powder, solutions, emulsions.
Supplier: Joel, F. W.
Polymethyl methacrylate; two-part, cold-setting material, (see N.H.P.)
Polyvinyl acetate; solutions (adhesives, e.g. UHU).
Supplier: Joel, F. W.; hardware, stationery and model shops.
Polyvinyl acetate; emulsions (e.g. Vinamul N 9146 and Evostik Resin W)
Supplier: Joel, F. W.; hardware, stationery and model stores; Sealjoint Ltd.
Polyurethane foam, two-part cold-setting systems.
Supplier: Strand Glass Co. Ltd.
Reprosil, dental moulding liquids, pastes and putty (see Silicone, cold cure silastomers).
Respirators.
Supplier: Joel, F. W.; some ironmongers, laboratory suppliers.
Rubber latex, various grades, fillers, pigments.

Supplier: Bellman, Ivey & Carter Ltd; Strand Glass Co. Ltd; Trylon Ltd.

Sieves.

Supplier: Hardware and ironmongery stores; Endecotts (filters) Ltd; laboratory suppliers.

Silicone, cold-cure silastomers (moulding compounds).

Manufacturer: Dow-Corning (for Silastics); I.C.I. Ltd (for Silcosets); De Trey's (for Reprosils).

Supplier: Claudius Ash & Sons Co. Ltd; Cottrell & Co.; Strand Glass Ltd; Tiranti, Alec. Ltd; Trylon Ltd; Hopkin & Williams Ltd.

Silicone fluid, e.g. M.S. 200/350 cs.

Supplier: Hopkin & Williams, Ltd.

Solvents and other chemicals.

Supplier: Joel, F. W.; laboratory suppliers; some pharmacies.

Vibro-Pen P2-X (pneumatic engraver).

Manufacturer and supplier: Desoutter Bros. Ltd.

Vibro-tool (electric engraver).

Manufacturer and supplier: Burgess Power Tools Ltd.

Vinalak 5911. Manufacturer: Vinyl Products Ltd.

Supplier: Check with manufacturer.

Addresses

Armstrong Cork Co. Ltd, 351 Caledonian Road, London N1.
Ash, Claudius & Sons Co. Ltd, 26 Broadwick Street, London W1.
Bellman, Ivey & Carter Ltd, 358b Grand Drive, West Wimbledon, London SW20.
Bottley, Gregory & Lloyd, 8–12 Rickett Street, London SW6 1RU.
Burgess Power Tools Ltd, Sapcote, Leicestershire, LE9 6JW.
Cottrell & Co., 15 Charlotte Street, London W1.
Courtin, Marcel, Courtin House, Horsham Road, Capel, Surrey.
Desoutter Bros. Ltd, The Hyde, Hendon, London NW9.
Dow Corning Ltd, Reading Bridge House, Reading, Berkshire.
(E.L.E.) Engineering Laboratory Equipment Ltd, Eastman Way, Hemel Hempstead, Hertfordshire, HP2 7HB.
Endecotts Ltd, 9 Lombard Road, London SW19.
Gallenkamp Ltd, PO Box 290, Technico House, Christopher Street, London EC2P 2ER.
GEC Mechanical Automation Ltd, Birch Walk, Erith, Kent, DA8 1QH.
Geological Museum (Bookshop), Exhibition Road, London SW7 2DE.
Griffin & George Ltd, Head Office, Ealing Road, Alperton, Wembley, Middlesex, HA0 1H5.
Hopkin & Williams, Ltd, Freshwater Road, Chadwell Heath, Essex.
Howmedica International Ltd, 622 Western Avenue, London W3 0TF.
(I.C.I.) Imperial Chemical Industries Ltd, Head Office, Millbank, London, SW1.
I.C.I. (Plastics Division), Bessemer Road, Welwyn Garden City, Hertfordshire.
Joel, F. W. Ltd, Museum, Laboratory and Archaeological Supplies, Unit 5, Old Medow Rd, Hardwick Industrial Estate, King's Lynn, Norfolk, PE30 4HH.
Morris, B. O. Ltd, Briton Road, Coventry, CV2 4LG.
Monsanto Chemicals Ltd (UK), Monsanto House, Victoria Street, London SW1.

Park Naval Engineering Ltd, Ordinance Street, Blackburn, Lancashire, BB1 3BN.

Scott Bader Ltd, Wollaston, Wellingborough Northants.

Strand Glass Ltd, (Head Office), Brentway Trading Estate, Brentford, Middlesex.

Tiranti, Alec. Ltd, 70 High Street, Theale, Berkshire.

Trylon Ltd, Thrift Street, Woolaston, Wellingborough, Northants.

Vinyl Products Ltd, Mill Lane, Carshalton, Surrey.

Index

abrasives, 44
acid preparation, 44–45
aluminium foil, 37, 38
ammonia,
  setting plaster of Paris, 36
  pyrite conservation, 48
aragonite, 30

bone, 30, 32–38
  materials for rapid removal of, 37–38
  pretreatment of in soft sediments, 32
  removal from hard rocks, 38
  reptile, 43
Butvar, 38, 39

calcite, 18, 30
carbonate, 44
casts, types of, 29–30
cataloguing, 23–24
cements, types of, 46–47
cephalopods, 43
chalk, precipitated, 50
chelating agents, 44
chemicals,
  disposal of, 51–52
  handling of in preparation and conservation, 50–51
chitin, 44
citric acid crystals, in setting of plaster of Paris, 35
cleaning,
  mechanical, 39–44
  chemical, 44–45
cocoon,
  plaster of Paris, 32–37
  polyurethane foam, 37–39
collection, arrangement of, 25–26
compression, 32
concretions, 30
conodont, 30, 31
consolidation agents, 38, 39, 45–46
cotton wool, 23, 25, 49

dam, in removal of fossil bone, 38
dental equipment, 41–43
diatoms, 43

echinoderms, chalk, 43–44
emulsion (Vinamul N9 164), 24
equipment,
  in extraction of fossils, 28–39
  in field, 8–14
  in mechanical cleaning, 39–43
expanded polystyrene, 28

filling material, types of, 47–48
foraminifera, 30, 31
fossils,
  in cleavage and bedding planes, 30
  friable, fragile, treatment of, 45

glass-fibre, 38, 39
glues, types of, 46–47

hydrochloric acid, for use in field, 14

invertebrate, 30
iron oxides, 44

labelling,
  in field, 27
  method of, 19–20, 25
  types of, 19–21, 23–24
limestone, 27, 30–32, 44

macrofossil, technique for recovery, 44
microfossil,
  sampling for, 31
  technique for recovery, 44
microscope, 31, 43

minerals,
  blooms, 18
  weathering, 18

nodules, 30–31
  ironstone, splitting technique, 45

ostracods, 31, 43

packaging materials, 21–22, 32, 49
palaeoecological records, 27
phosphatic fossils, chemical cleaning of, 44
plaster of Paris, 32–37
polyethylene glycol, 28–29, 42
polyurethane foam, see cocoon,
polyvinyl acetate, 24
polyvinyl alcohol, 39
preservation, 22–23
pretreatment, 28–29
protective clothing,
  in field, 8–9
  in laboratory, 44, 51–52
pyrite, 44, 48
  prevention of decay, 50

radiolaria, 30
recording, methods of, 19, 27
resin ribs, 38
root canal-PR-30, 43

salt solutions, 44
sandstone, cleaning of fossils in, 44
scrim, plasterer's, 34, 35
seccotine, 23
shellac, 28
silica, 30
silica gel, 50
silicone fluid, for fossils preserved as pyrite, 50
slaty rocks, 30
splints, 35, 36, 38–39
sponges, 30
squeezes, 29
storage, 24–26, 49–50

talcum powder, 30
thioglycollic acid, 44
tissue paper, 32, 49
  acid free, 25
tools, see equipment,
toxic fumes, during preparation of fossils, 38

ultrasonic probes, 43
ultrasonic tanks, 43
undercuts, 29, 32
urethane, 38

ventilation, 38, 50, 51
vertebrates,
  pretreatment of bones, 28–31
  pretreatment of large vertebrates, 28
  pretreatment of small articulated vertebrates, 28

weathering, 30

zinc sulphate, setting plaster of Paris, 36